INTRO2CRYPTO

LLC EDITION

By Brian Scott

Copyright © 2021 by Brian Scott

All rights reserved.

No part of this book may be reproduced or used in any manner without the prior written permission of the copyright owner, except for the use of brief quotations in a book review. To request permissions, contact the publisher at intro2cryptollc@gmail.com

First paperback edition July 2021

ISBN 978-1-7369552-3-9 (paperback)

ISBN 978-1-7369552-4-6 (ebook)

Dedication

I dedicate this edition to my mom Joyce "Douthet" Spain. We hardly ever see eye to eye, but I love you very much. The past is the past and I love how the future looks with your grandson Lance, you and me.

Love your son,
Brian Scott

Now that Lance was the owner of two Bitcoin ATMs, he wanted to get the LLC as soon as possible. When he got home he began to research how to get an LLC. Lance saw several different prices. He decided to try to apply for the LLC himself.

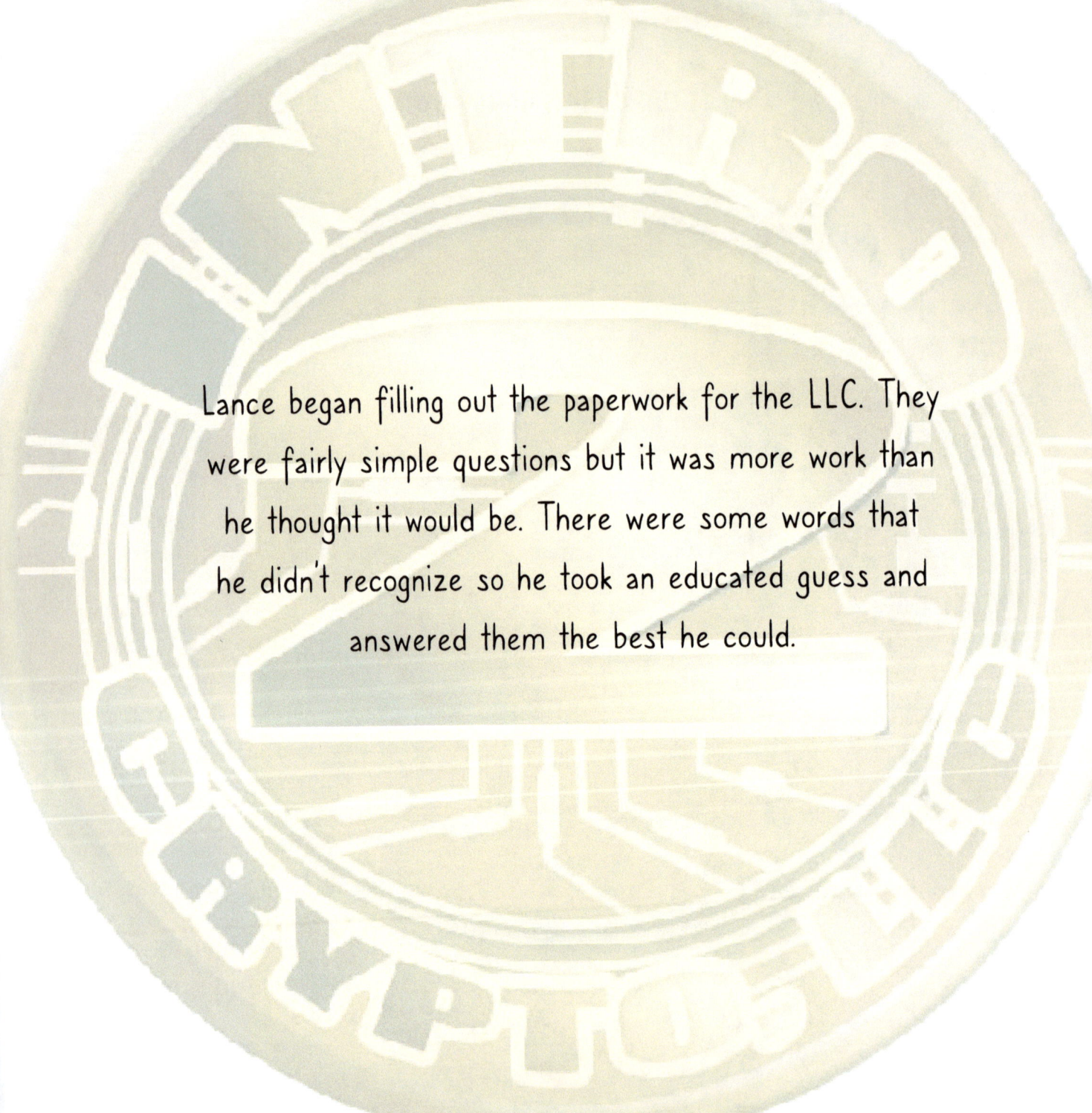

Lance began filling out the paperwork for the LLC. They were fairly simple questions but it was more work than he thought it would be. There were some words that he didn't recognize so he took an educated guess and answered them the best he could.

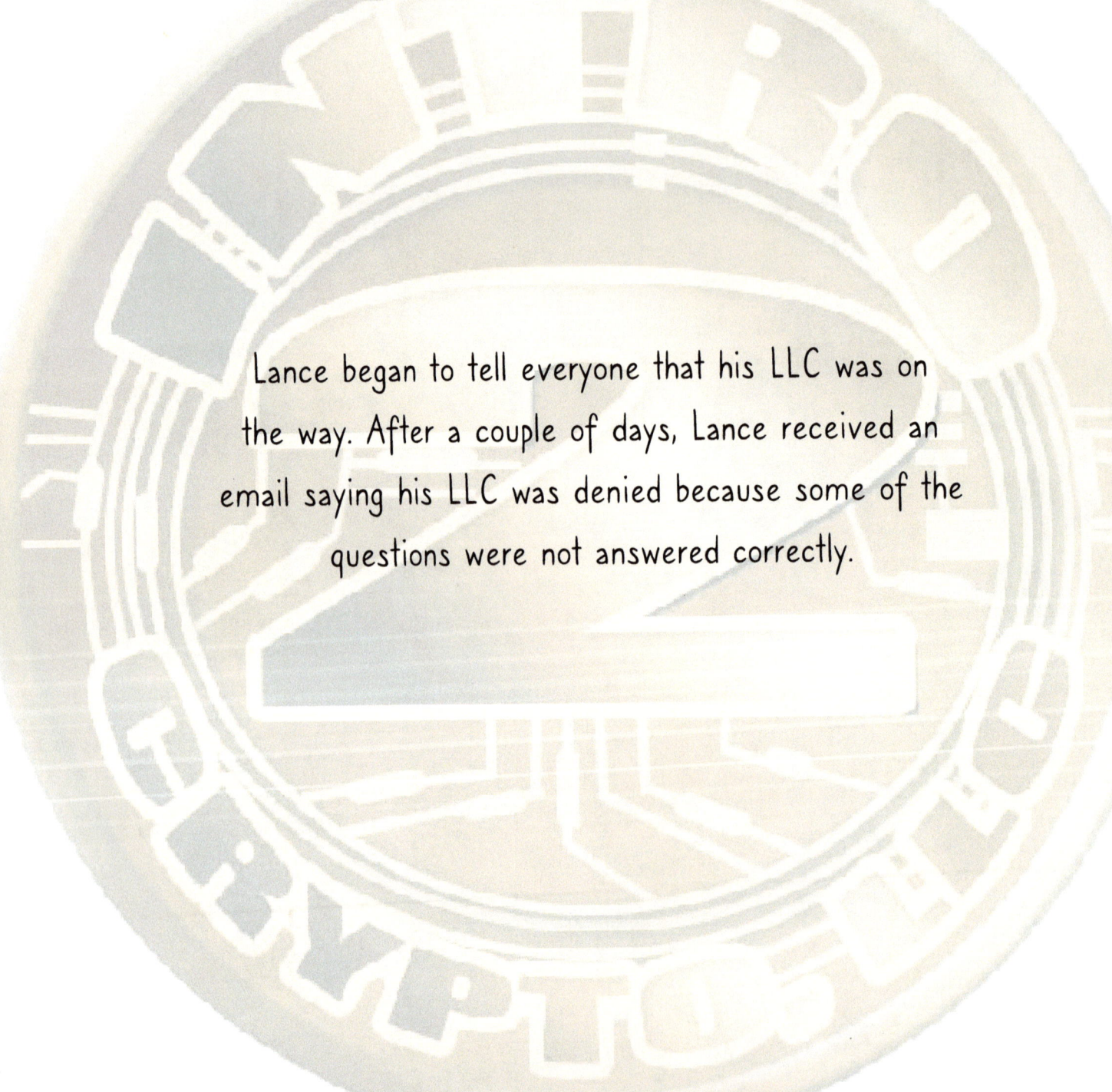

Lance began to tell everyone that his LLC was on the way. After a couple of days, Lance received an email saying his LLC was denied because some of the questions were not answered correctly.

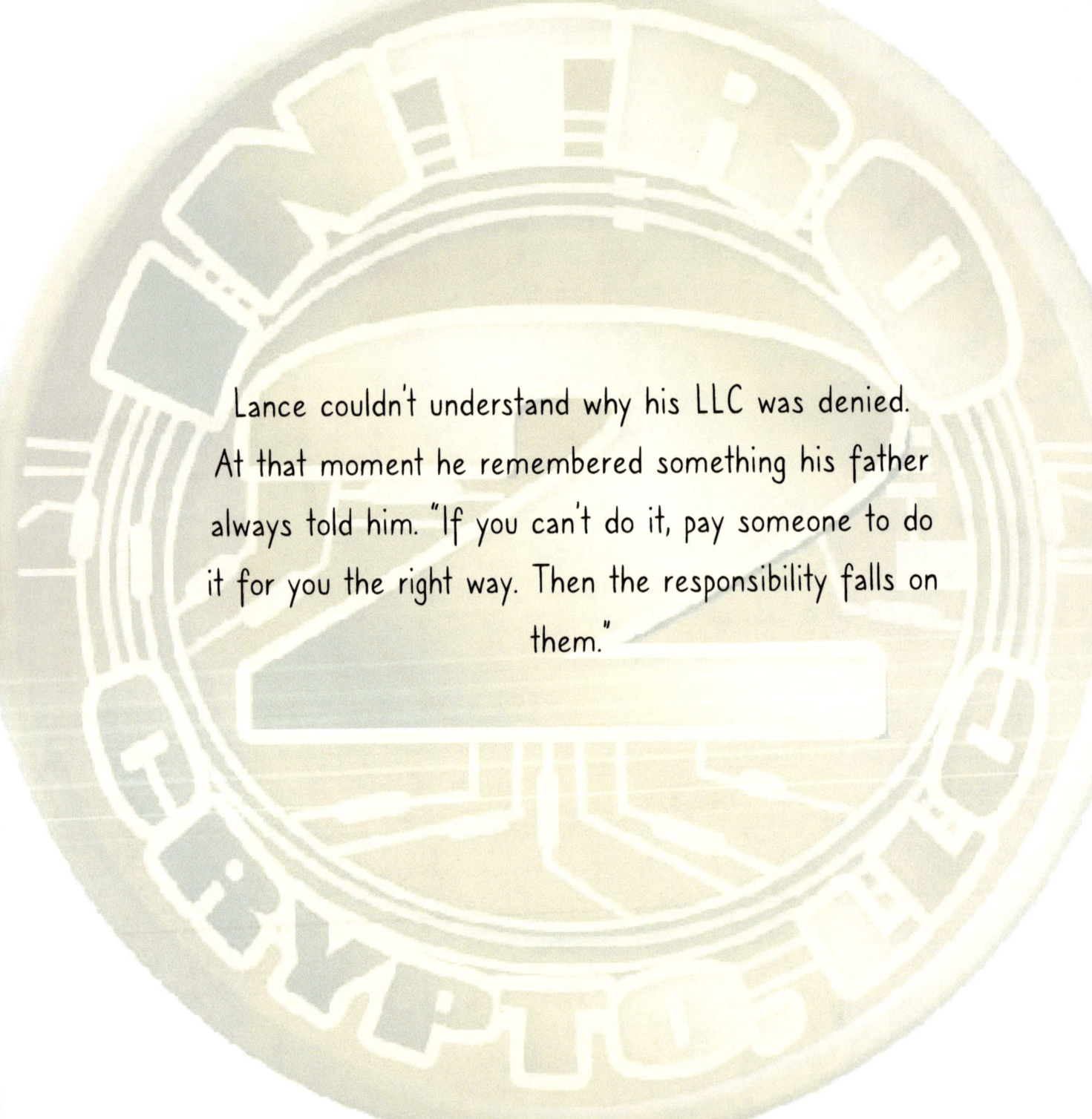

Lance couldn't understand why his LLC was denied. At that moment he remembered something his father always told him. "If you can't do it, pay someone to do it for you the right way. Then the responsibility falls on them."

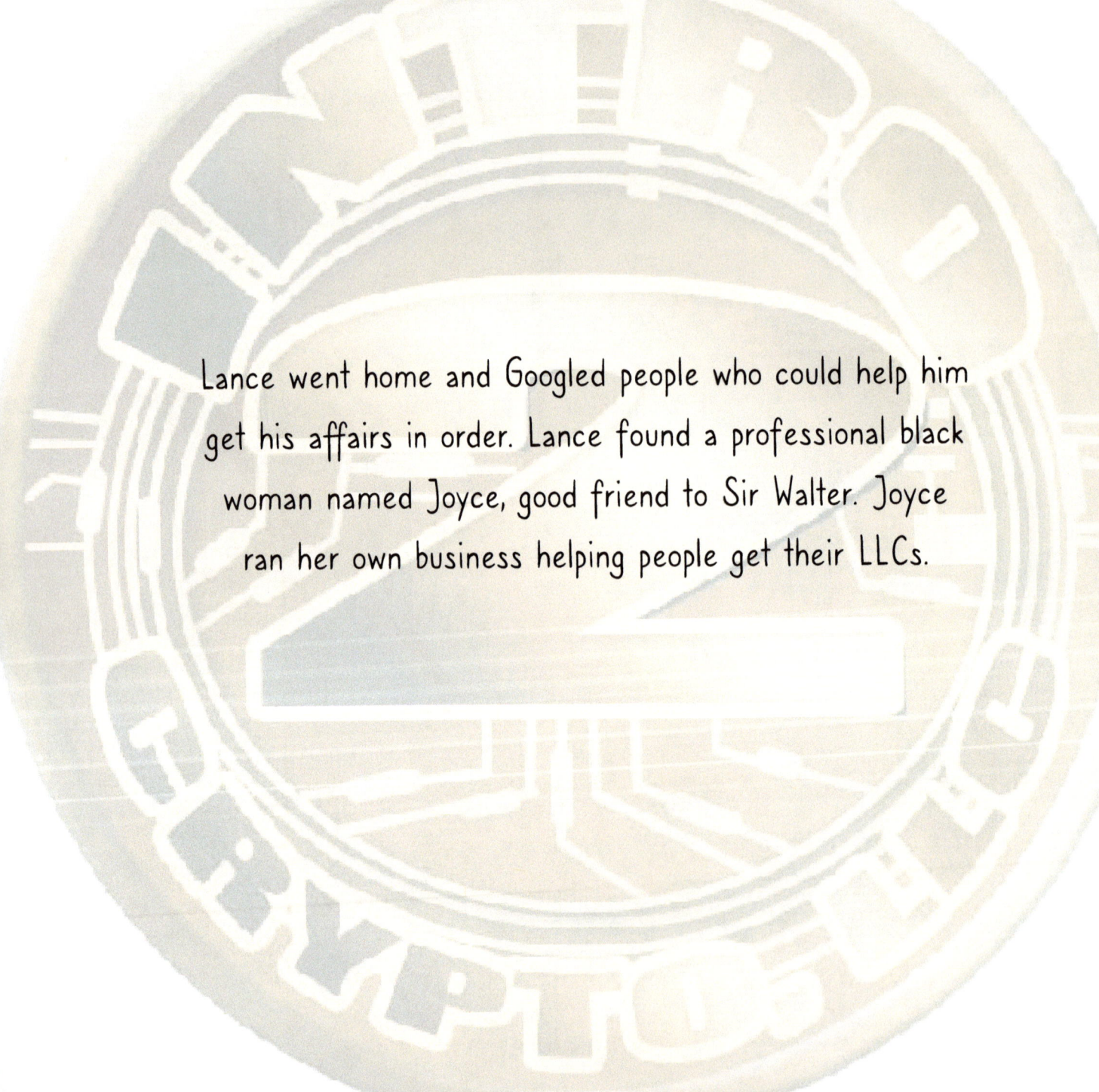

Lance went home and Googled people who could help him get his affairs in order. Lance found a professional black woman named Joyce, good friend to Sir Walter. Joyce ran her own business helping people get their LLCs.

When Lance sat down in her office he began to discuss why he was getting the LLC for his Bitcoin ATM. Lance has one Bitcoin ATM in the barbershop and he needs a location for his other one.

Joyce began to ask questions about Bitcoin. How does it work? What is the value? What makes it so valuable? Isn't cash better? Lance began to explain all these things to Joyce. Joyce was surprised by the depth of knowledge Lance had about Bitcoin and its longevity of investments.

When Lance was done explaining everything Joyce asked, "How can I purchase crypto?" Lance replied, "You can own bitcoin today." Lance offered to pay Joyce in Bitcoin for the LLC. Joyce was very excited and offered Lance a new location to put his Bitcoin ATM. Lance replied "That's a deal. One hand washes the other sista!!"

Joyce told Lance to finalize that he needs to name the LLC. Lance came up with the "Bitcoin Withdraw, LLC" Joyce thought the name was very intriguing and wanted to know more. Lance explained to Joyce how Bitcoin is a learning process and you have to continue to learn. Once you learn it can never be taken away from you.

In order for Lance to protect his investment in Bitcoin, he has to study and keep up to date with the latest technology. Lance heard about a big workshop being held in Dubai. Dubai is known for its use and knowledge of this technology. He then seeks out his quest for knowledge and books a flight to Dubai.

TO BE CONTINUED..........

Other Children's Books by Brian Scott

Intro2Crypto Kids 1st Edition

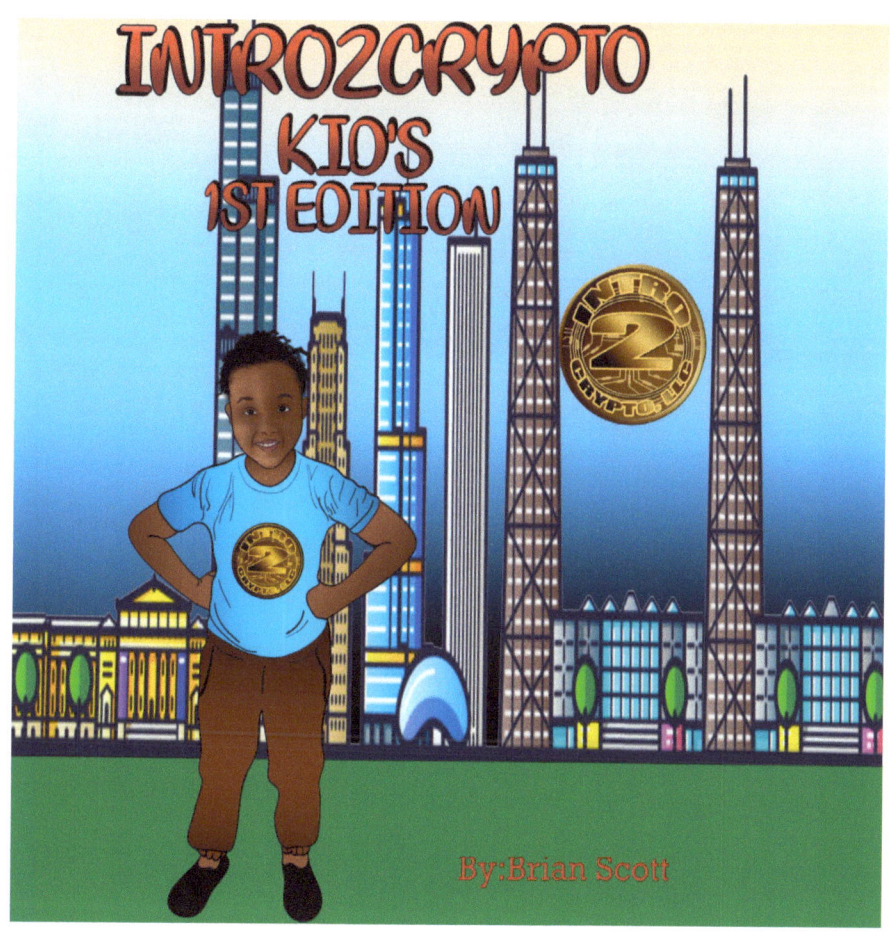

Intro2Crypto Bitcoin ATM Edition

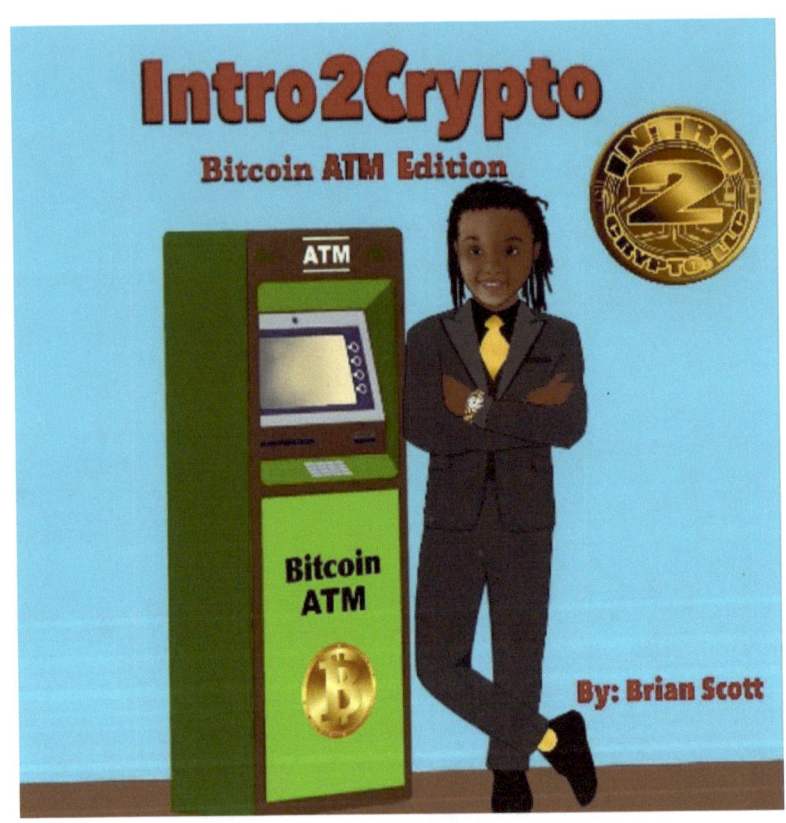

www.ingramcontent.com/pod-product-compliance
Lightning Source LLC
Chambersburg PA
CBHW040032050426
42453CB00002B/85